The Amazon Rainforest

By Elizabeth Theurnagle

Cover design by Robin Fight

goodandbeautiful.com

The Amazon Rainforest

Come with me on a journey through one of the most exotic places on Earth—the Amazon Rainforest. Things that may seem unusual where you live are actually quite normal here.

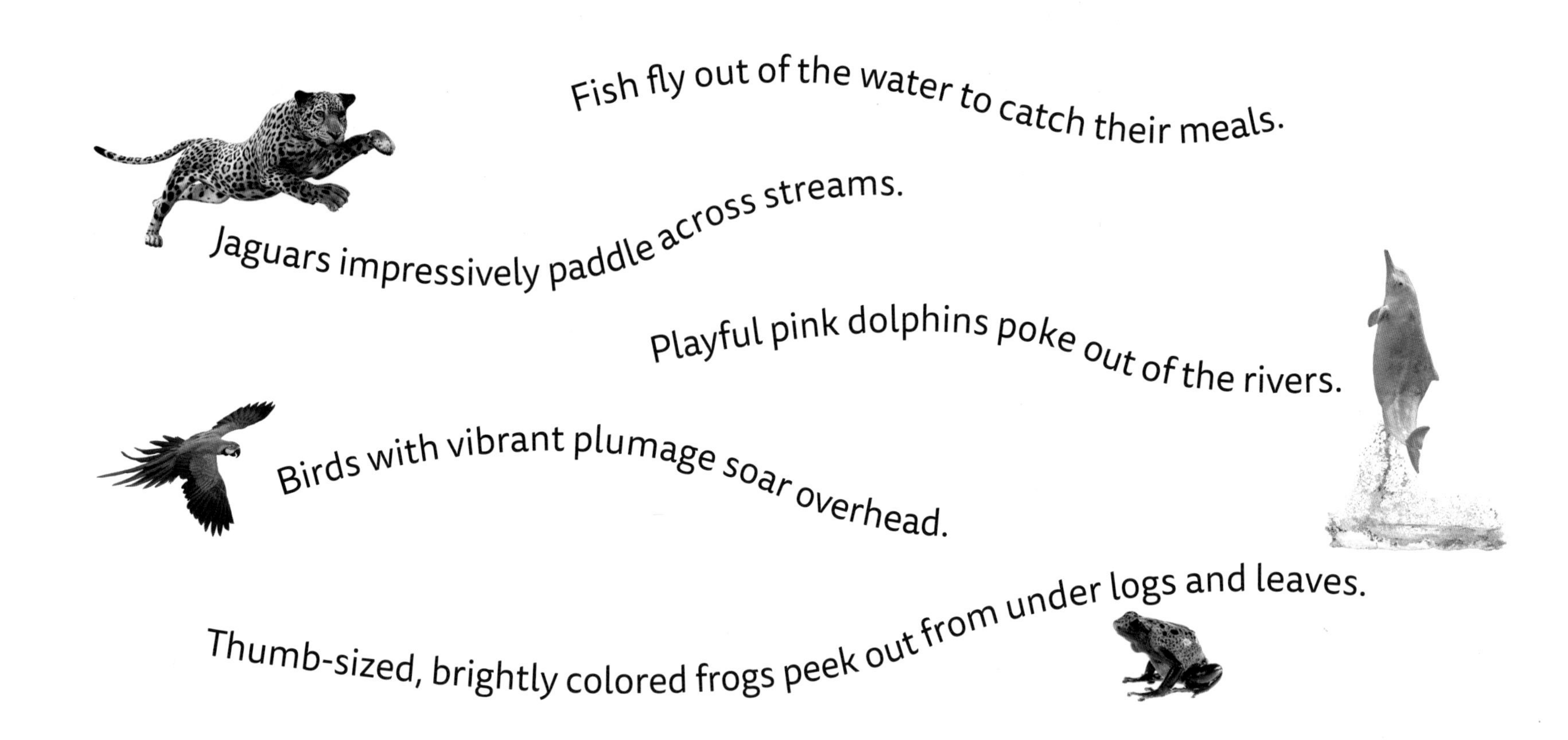

From the bottom to the tip-tops of the tallest trees, the dense jungle of the Amazon is a battle zone for survival. Animals compete for food and a mate. Plants have bizarre yet beautiful features that allow them to thrive.

The hot, humid air feels steamy—like you just stepped out of a hot shower. It is December, and the smell of an approaching storm signals the beginning of the rainy season. This can last months, so put on your raincoat and rain boots before we move on.

The Amazon Rainforest is massive—about the size of the continental United States. It is shared by nine South American countries: Brazil, Peru, Colombia, Bolivia, Venezuela, Ecuador, Guyana, Suriname, and French Guiana. It spans from the Andes Mountains on the west all the way to the Atlantic Ocean on the east.

A rainforest needs three important things: rain, soil, and sunlight. The Amazon receives plenty of those things. The Amazon Rainforest is between the two tropics: the Tropic of Capricorn and the Tropic of Cancer. This makes the Amazon a tropical rainforest.

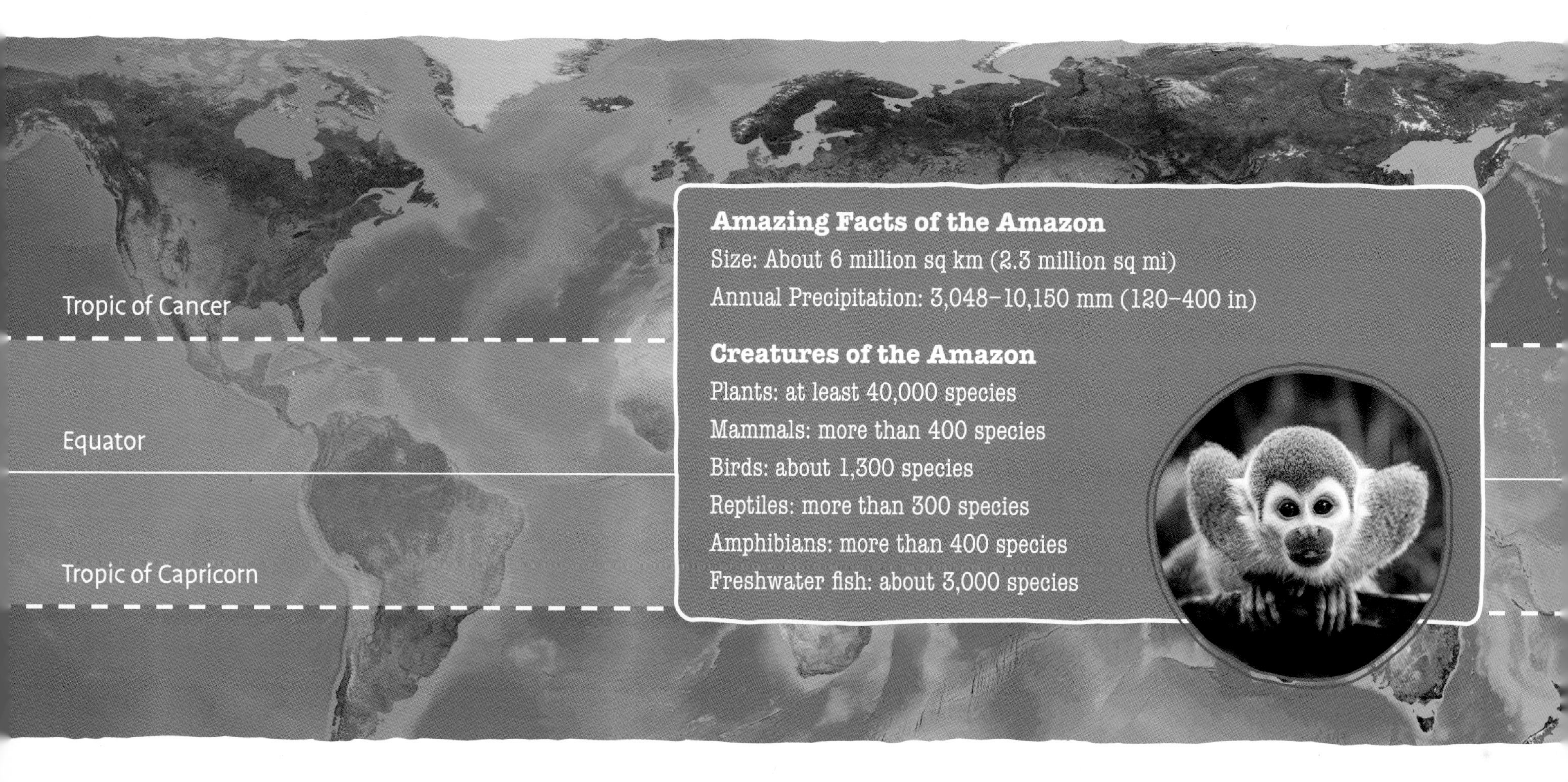

Amazing Facts of the Amazon

Size: About 6 million sq km (2.3 million sq mi)

Annual Precipitation: 3,048–10,150 mm (120–400 in)

Creatures of the Amazon

Plants: at least 40,000 species

Mammals: more than 400 species

Birds: about 1,300 species

Reptiles: more than 300 species

Amphibians: more than 400 species

Freshwater fish: about 3,000 species

species (n.)

Animals of different types are called different **species**. For example, a hawk and a parakeet are different species of birds.

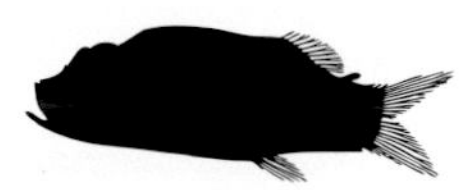

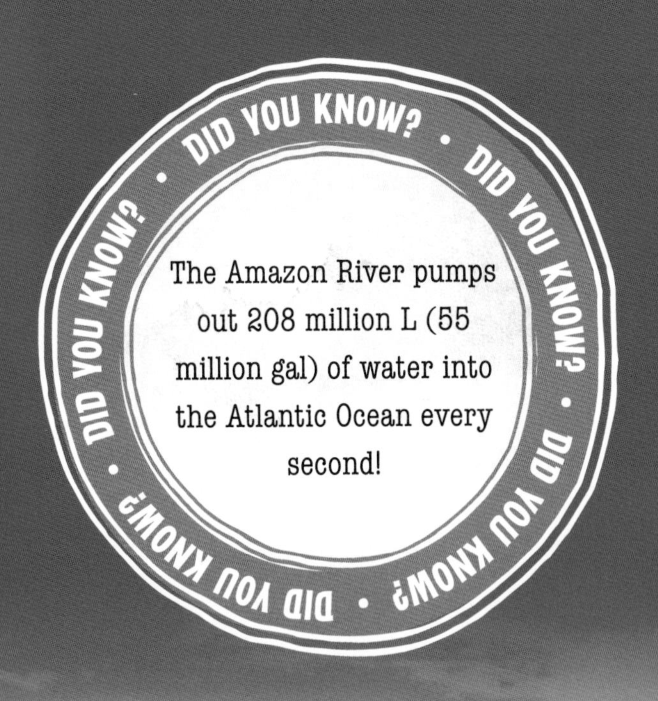

Tropical rainforests only cover about 6% of the earth, but they provide homes for more than half of the plant and animal species in the world. Life-giving rain creates a rich habitat for millions of plants, animals, and other wildlife—some of which are very rare.

Where does all this water come from? Running through the Amazon Rainforest toward the Atlantic Ocean is the shimmering Amazon River. If you could swim from one end of the river to the other, you would travel about 6,400 km (4,000 mi). That is exactly what a man named Martin Strel did. It took him sixty-six days! Once you reach the Atlantic Ocean, you would be surprised to discover that a large amount of this fresh river water continues to flow into the ocean. You could even swim another 200 km (125 mi) into the Atlantic Ocean before you would reach saltwater.

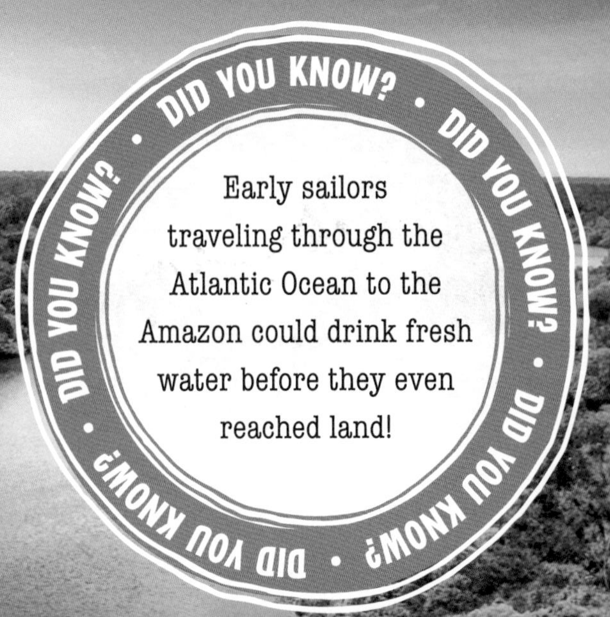

Where does the Amazon River get all this water from? It is fed by more than 1,100 tributaries that flow into it! These smaller rivers and streams receive water as snow melts from the nearby Andes Mountains. With this large supply of water, the Amazon Rainforest is self-sustaining. This means it can take care of itself as the water provides for all life in the rainforest.

Water within the leaves of the Amazonian plants transpires (leaves the plants as water vapor) and collects as low-level misty clouds. This vaporized water, in addition to water that collects in clouds through evaporation, eventually falls back to the earth as rain—giving and receiving water all within the Amazon.

Where does the largest rainforest (the Amazon) get fresh soil? This answer may surprise you—all the way from the Sahara Desert on the continent of Africa! As Saharan sand blows across the Atlantic, it drops in the Amazon, adding fresh nutrients to the soil.

Plants produce enough oxygen to support life in the rainforest, and there is enough plant and animal life to be consumed as food to support all plant and animal life, thus making the Amazon a self-sustaining ecosystem.

Clouds form above the Amazon River Basin through evaporation and transpiration.

Amazonian Seasons

The Amazon doesn't have the four seasons of fall, winter, spring, and summer. Instead, it has two seasons—rainy and dry. During the months of the rainy season, clouds release huge amounts of rain. In addition, water from the melting snow in the Andes runs down the mountains into the tributaries, which feed into the Amazon River. The riverbed can't hold all of this water, so it floods. The flood zone of the forest floor becomes completely submerged in water. Land-dwelling animals scamper up trees and scramble to higher ground for safety. Plants that were once homes for flying and crawling creatures now shelter water animals such as fish.

DID YOU KNOW?

The Amazon can receive 2.5–5 cm (1–2 in) of rainfall within only one hour.

river basin (n.)

The area of land that has tributaries flowing toward the main Amazon River forms a **river basin**.

Families that live in the rainforest build their homes on stilts. When the floods come during the rainy season, their homes stay above the water level.

A silver arowana hovers right below the water's surface and patiently waits. When an unsuspecting insect gets close by, the fish leaps out of the water. Its jump can reach up to 3 m (9.8 ft)! Because of this, the silver arowana has earned the nickname "water monkey."

From above, the Amazon Rainforest may simply look like a sea of trees. However, hidden under the canopy of leaves lays an entire network of plants, insects, mammals, amphibians, reptiles, and other living things. The Amazon is home to the largest rainforest ecosystem in the world. With so much biodiversity, much of the ecology remains unknown. Some creatures native to the Amazon are not found anywhere else in the world, and some are so well hidden they have yet to be discovered.

Clouds over the Amazon

Every species in the rainforest, from the tiny ant to the large jaguar, is important. In order to survive, all forms of life need each other and the water, air, and other resources in their environment. Several food chains form a massive food web.

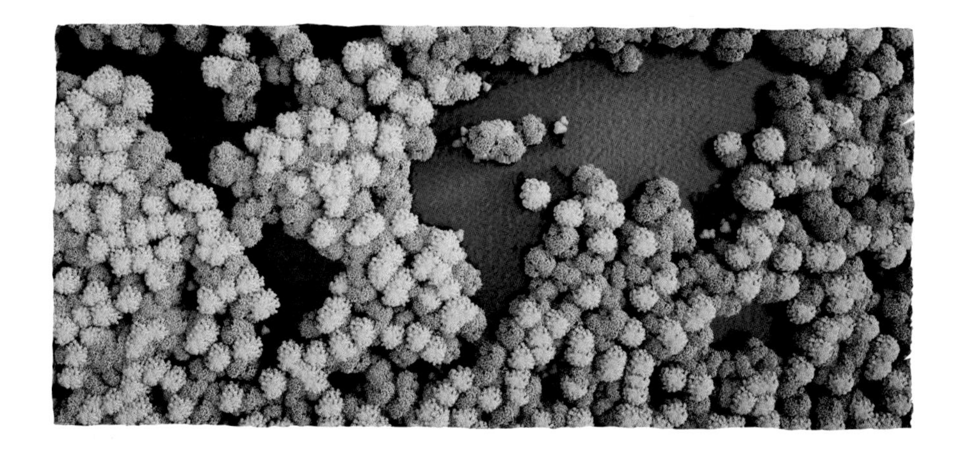

Creatures within an ecosystem depend on one another and their environment. Brazil nuts are eaten by an agouti (a rodent). Because of their exceptionally strong teeth, they can chew through the tough covering. The nuts provide nutrients for the agouti, and the agouti spreads the seeds of the Brazil nut tree for more to grow.

food chain (n.)

A **food chain** shows what eats what in an ecosystem.

food web (n.)

Many food chains together form a **food web**.

A plant's or animal's main role in the ecosystem tells you if its job is a producer, consumer, or decomposer. Grab your magnifying glass and your binoculars, and let's find each of these roles in the rainforest.

Many varieties of plants are all around you. Tall trees reach the heavens while shorter shrubs don't grow very far from the ground. Some leaves are narrow and long, and others are giant and wide. Certain types of plants have large roots that anchor them to the ground, while some types never touch the ground. Instead, they live on the branches of other plants. Regardless of how they look, plants are producers because they "produce" their own food from sunlight. They do this through photosynthesis, a process performed within their leaves to convert sunlight into sugar, which is used for energy.

As you walk through the Amazon, you kick up some dead leaves and twigs. You inspect the ground with your magnifying glass, and you see a colony of termites! These little critters are decomposers and are very important to the rainforest ecosystem. They eat dead plant matter such as leaves and twigs. Their waste replenishes the ground with nutrients for growing plants to use. Termite burrows create tunnels for rainwater to seep through. Without termites and other decomposers, such as ants, worms, millipedes, and bacteria, the rainforest could not survive. It would get littered with dead plants and animals with no way to break them down.

Flying overhead you spot a beautiful toucan. You pull out your binoculars for a better look. This large-beaked bird perches on a sturdy branch. Using its strong beak, it reaches for a fig hanging off the tree. The edges of its beak are serrated like a saw, allowing it to peel open fruit. All consumers, like this toucan, eat plants or other animals. This toucan's favorite foods are fruit and insects.

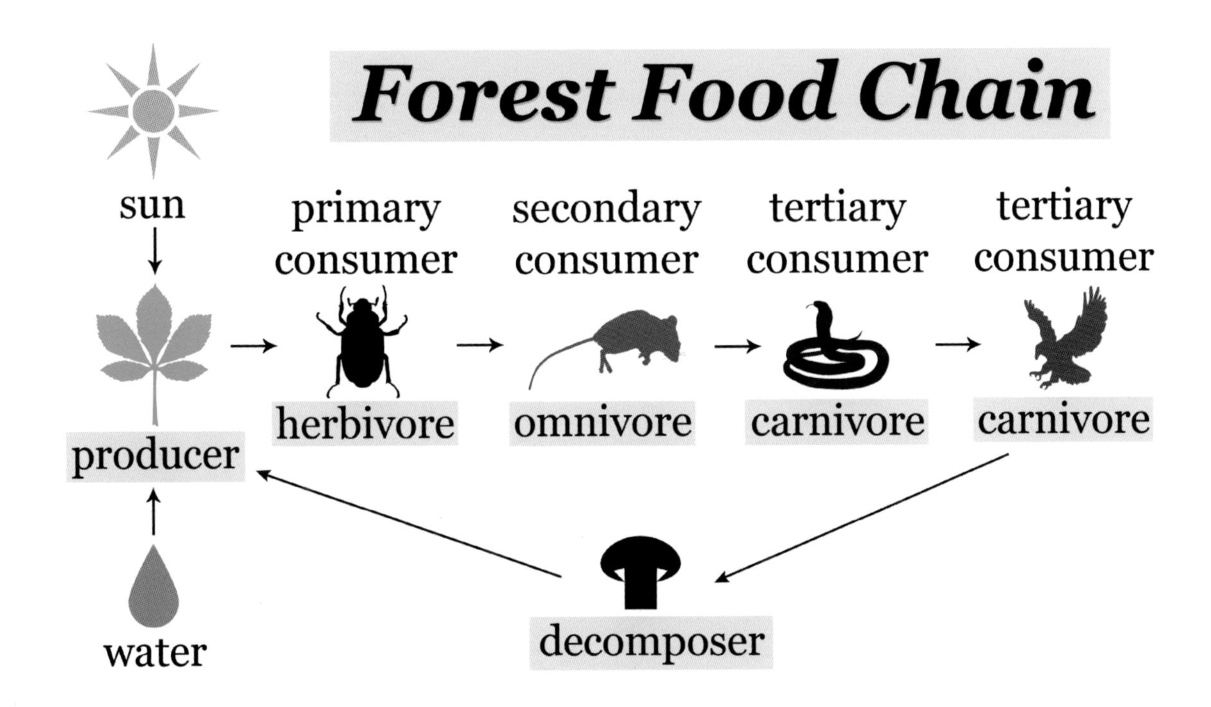

WHAT DO YOU THINK?

Consumers that eat just plants are called herbivores. Consumers that eat either plants or meat are called omnivores. Consumers that mostly eat meat are called carnivores. What do you think insectivores and frugivores mostly eat?

Multiple food chains within an ecosystem form a food web. Energy is transferred from the producer to the consumer. When producers or consumers die, energy is transferred to the decomposer, which restores nutrients into the ground, which once again helps producers grow.

Up ahead you see a shadowy figure lurking in the dim underbrush. You quickly hide behind a tree and slip on a camouflaged poncho. The figure steps forward, and you discover it's a jaguar! These sleek creatures depend on their stealth and speed to survive. As hunters of the forest, they prey on smaller consumers such as small monkeys, birds, armadillos, and capybaras. Their ability to swim also allows them to fish. Like your poncho does for you, the pattern of spots on jaguars' coats provides camouflage for them. They carefully stalk their prey, and if they aren't sneaky enough, their dinner will discover them and quickly run away.

Jaguars sit at the very top of the food chain and are called apex predators. They are the toughest consumers. Other big cats, such as leopards, and fierce reptiles, such as the green anaconda and the caiman, are also apex predators. Being at the top of the food chain is not easy. Apex predators compete with each other for food. They are always on alert and must stay strong and healthy.

Lower-level consumers also compete as they must find ways to outsmart other animals. Bats and frogs eat the same kind of fruit that insects and birds eat. To make sure there's food for bats and frogs, and to avoid getting snatched up and becoming a meal for another predator, they've learned to hunt at night. They'll feast on fruit and bugs that the birds and insects didn't finish during the day.

camouflage (n.)

Camouflage allows a creature to blend in with its environment.

Apex Predator: Harpy Eagle

As the ruler of the rainforest canopy, this large eagle has keen eyesight and powerful wings. Its strong talons can pick up prey more than half its weight. Rainforest creatures screech to sound an alarm when the harpy eagle is spotted—signaling "take cover!"

Apex Predator: Black Caiman

This fearsome predator weighs about 360 kg (800 lbs) and lives in the rivers of the Amazon. With eyes and nostrils strategically placed on the top of its head, it can hide in shallow waters for a long time until an unsuspecting victim gets too close.

Apex Predator: Green Anaconda

Known as the longest and heaviest snake in the world, the anaconda doesn't use venom to kill as other snakes do. Instead, it wraps around its prey to constrict breathing then swallows it whole. They have even been known to kill jaguars!

WHAT DO YOU THINK? • WHAT DO YOU THINK? • WHAT DO YOU THINK?

An ecosystem is a network of living (plants, animals, and insects) and non-living (water, air, and climate) things within an environment that depend on one another to survive and affect each other.

- What do you think would happen to plants and animals if the climate changed so that it no longer rained as much?
- What do you think would happen to top consumers, such as apex predators, if disease caused a whole host of smaller mammals to die?

Year after year, as the floods come and go, more than a rainforest is sustained. Swamps, savannas, and villages are part of the Amazon basin.

When the rainy season is over, the rain slows enough to allow animals to return to the parts of the land that were once flooded. Take off your raincoat and grab some bug spray! We are on our way to explore the layers of the rainforest.

Amazon Water Lily

As we shift to warmer, drier days and set our hearts upon exploring the rainforest layers, we pass by a unique plant that piques our attention. Resting on pools of water is the Amazon water lily (*Victoria amazonica*). Their lily pad leaves are gigantic and strong. A leaf can extend more than 2.5 m (8 ft) in diameter and can hold so much weight a person can sit on it!

Nightfall signals a magnificent change in the flower of this beautiful plant. At dusk, the white flower's temperature increases, causing it to open. In crawls a pollinator beetle. While the beetle feasts on the tasty nectar and pollinates the flower, the flower closes with the beetle still inside. From this time to the next evening, the flower petals change from white to pink. It then opens its petals and releases the beetle. The beetle flies off to repeat this process, but the flower has ended its short life cycle and turns to seed.

With feet planted firmly on the ground, you look high into the sky and notice that your view is blocked by a canopy of draping leaves. The rainforest can be divided into four layers: the forest floor, the understory, the canopy, and the emergent layer.

A single tree can shoot through all four layers as its roots begin on the floor and its topmost branches extend through the highest limit. Several trees can house entire ecosystems—insects and vines that live around the trunks, agile monkeys and slow sloths that reside in the leafy branches, and the bold birds of prey that guard their nests at the highest treetops.

We'll begin the exploration of the forest layers right where we stand—on the forest floor.

The Forest Floor

The forest floor is carpeted with dead and freshly fallen leaf litter, old twigs, nuts and fruits that have dropped from above, and busy decomposers. Vines twine about, climbing up tree trunks, and new saplings desperately search for a glimmer of light to peek through an opening in the canopy overhead.

It feels damp and hot, and despite the fact that it's the middle of the day, the forest floor is very dim. Only around 2% of sunlight reaches the ground because the light gets filtered out by the umbrella of leaves in the canopy. Occasionally, when a large tree falls, a light gap opens up. Seedlings on the floor that were once dormant (asleep) sense the presence of light and begin to rapidly grow. These seedlings race to become the next new tree to reach the canopy.

While the busy sounds of birds chirping and monkeys chattering come from up above, the forest floor is no idle place. As seeds and fruits fall from the tops of the trees, small rodents gather them for food. Decomposers such as termites, cup-shaped fungi, and bacteria dutifully break down old leaves and dead twigs. Small animals such as centipedes, crab spiders, woodlice, snails, and worms scavenge about, eating the remains of decaying life.

Large-leafed shrubs provide perfect hiding spots for large cats, as their coats blend in seamlessly, allowing them to spy on their prey.

Soil Ecology

Most of the forest floor soil in the Amazon is low in nutrients. Unlike temperate rainforests that have rich soil, most of the nutrients in tropical rainforests come directly from decaying matter that is broken down quickly and used up almost immediately. These wastes can be found enveloped by butterflies or rolled into balls and taken away by dung beetles. Simply put, the nutrients are recycled so fast that there isn't enough time for them to fully seep into the soil. Therefore, the soil remains poor in nutrients.

Leafcutter Ants

As you walk about, you notice a series of earthy mounds. Off in the distance, you spot a giant leaf with holes cut out. Upon closer inspection you see leafcutter ants diligently sawing away sections of leaf. Once pieces are successfully cut off, they are carried away. The ants, holding their leaf pieces like trophies, march along a distinct pathway back to their mounds. Small ants ride on the pieces of leaf, acting as guards against parasitic flies. Inside these mounds are underground mansions of many rooms and millions of ants.

These astounding ants cut more greenery from the rainforest than any other animal, removing about 12–17% of all leaves. Cutting through the leaves is possible because these dedicated leafcutter ants are equipped with razor-sharp jaws that vibrate at 1,000 times per second! Leaf pieces can weigh twenty times the weight of the ant carrying it.

Surprisingly, leafcutter ants do not eat the leaves. Rather, they chew them up into sticky clumps and store them in a room in their mound where they create a fungus that grows on their leaf clumps. A fungus garden forms, and it is the fungus that they feast upon.

Leafcutter ants are essential to the rainforest ecosystem. Their hard work of cutting up leaves prunes the forest and prompts new plant growth. Reducing leaf coverage also allows more light to enter the forest floor, and their fungus gardens enrich the soil.

Buttress Roots

Many of the trees you've walked by have tall, triangular, wall-like extensions coming out of the trunk. These sections of the trunk are actually the tree's roots! Buttress roots, as they are called, can reach as high as 4.5 m (15 ft). A buttress is an object used for support that is built against a wall. This is quite an appropriate name as these roots provide support for these giant trees. By comparison, trees that grow in temperate forests usually grow deep, strong roots that anchor them into the ground. Because the soil in a tropical rainforest such as the Amazon is nutrient-poor, rainforest tree roots are shallow. They grow outward in order to reach any nutrients, usually from decaying plant and animal matter, lying on the very top of the soil. However, shallow roots provide little support, and a tree can easily fall over. To overcome that weakness, buttress roots sturdily surround the central focus of the tree, holding it up. Therefore, buttress roots have two main purposes: to provide support and to access surface-level nutrients by reaching far and wide.

DID YOU KNOW?

Not all parts of the rainforest soil are poor. Areas that are part of the floodplain (the area that gets flooded during the rainy season) and edges of the forest will have more minerals in their soil.

Change into your pajamas because it's nighttime! However, it doesn't feel like the forest is going to sleep. Instead, the forest gets louder as nocturnal creatures awaken from their slumber—insects such as cicadas and grasshoppers chirp, frogs begin to croak, and even some mammals such as the nine-banded armadillo, tapir, honey bear, owl monkey, and a petite wildcat known as the ocelot begin their day . . .or, rather, their night!

Nocturnal Creatures of the Amazon

Owl Monkey

The large round eyes of owl monkeys help them see well in lower lighting, making them much better at seeing in the dark than any other primate. A typical day for owl monkeys includes waking up shortly after sunset and then snacking on insects by plucking them off a branch or snatching them out of the air as they fly by. Owl monkeys group together to forage for fruit, flowers, leaves, and nectar. Before sunrise they return to their sleeping site or find another one close to home. These include hollows in trees or under the foliage of plants that grow on tree branches. Sometimes they'll even share their bed with other nocturnal animals such as bats.

Honey Bear

Also called kinkajous, these fuzzy mammals are not bears but are raccoon relatives with some monkey-like features. Their prehensile tails allow them to grip and hang from tree branches. They also have feet that can turn backward, allowing them to switch directions as they run up, down, or along limbs. Sharp claws help the honey bear dig in to eat fruit and smaller mammals. As the name suggests, honey is a favorite treat. Their anteater-like tongues enable them to access honey from a hive and are also useful in snagging a few termites and other insects from their mounds.

Ocelot

This small wildcat roams from the forest floor to the misty branches of the cloud layer at the top of the forest. As excellent swimmers and climbers, ocelots hunt practically whatever they can catch—mice, armadillos, monkeys, birds, reptiles, and during the rainy season, even fish or crabs. They are very patient hunters and are known to sit and wait as still as a statue for up to an hour at a time. Since they are primarily nocturnal, hunting is typically done at night or when it's cloudy. During the day they rest under large roots, bushes, or bundles of vines.

prehensile tail (n.)

A **prehensile tail** is able to grasp and hold things. It is especially useful for creatures living high up in trees.

The Understory

The next morning you awaken and climb up to the understory. This layer starts about 1.5–6 m (5–20 ft) from the forest floor. It's still warm and humid, and you're still under the ceiling of the leafy canopy. You come face to face with a leaf with legs, only to quickly realize that what you're looking at is not a leaf at all, but a katydid. These nocturnal insects are now going back to sleep for the day. Katydids use an ability called mimicry, where they copy (or mimic) the appearance of a leaf. This helps them blend in with their environment to reduce their chances of getting plucked off by a predator. Other creatures that use mimicry and camouflage to avoid being eaten are snakes, birds, tree frogs, lizards, jaguars, leopards, and some insects.

You feel water sprinkle your face as rain starts to filter down through the canopy, but there's still not much sunlight—only little patches of light here and there.

In the understory of the Amazon, you'll find shrubs with large leaves, which help the plant capture as much sunlight as possible. Vines that are rooted in the forest floor continue to make their way up the trunks of the trees. The full-grown trees found here are usually palms or woody plants.

Plants called epiphytes take root on tree branches and never experience the forest floor. Certain types of ferns, bromeliads, and orchids are examples of epiphytes and spend their lifetime on these tree branches.

This prime, shady spot of the understory keeps you sheltered from the wind, rain, and hot sun.

mimicry (n.)

Using **mimicry**, a creature can copy (mimic) the appearance of something else.

epiphyte (n.)

An **epiphyte** grows on another plant without harming it.

Poison Dart Frog

Some of the most vibrant colors on the earth can be found in the Amazon, and the frogs are no exception. The colors of poison dart frogs act as a warning to predators to stay away. Ingesting the poison of these frogs results in nausea, swelling, paralysis, or even death. There are about 200 species of poison dart frogs, but the title of most deadly goes to the golden poison dart frog. One of these frogs carries enough poison to kill ten human adults. Knowing the deadliness of the poison, native Amazonian hunters would lace their blow darts with dart frog poison to allow them to hunt more efficiently.

A mother dart frog lays her eggs, which will become fertilized by a male dart frog. Baby tadpoles begin to grow, and right away they are at risk for either drying out or becoming prey. To solve both of these concerns, the mother and father dart frogs take turns lying on the tadpoles to keep them moist. They also keep watch, because at this stage the tadpoles have not yet developed their toxins, but predators know not to come near the brightly colored frogs standing guard. Once the tadpoles hatch, they swim onto their father's back, and a thin layer of mucus keeps them in place. The father begins a treacherous journey up a tree looking for a safer place where he can set them so they can continue to develop. This safer place is usually in a pool of water within the leaves of a bromeliad plant (or any other plant with leaves that can catch and hold water). The tadpoles are deposited, and they continue to develop for about 6–12 weeks. During this time the tadpoles continue to receive nutrients from the water and from their mother, who brings them unfertilized eggs as a food source. Once they have matured, they gain their bright colors and poison. A dart frog's average lifespan is six years.

The Canopy

As you climb up higher into the layers of the rainforest, the rhythmic humming of insects fills your ears. Birds dart about, and monkeys glide from branch to branch. The canopy is teeming with life—snakes, sloths, butterflies, and bats are only a few of the creatures that call this layer home. Sitting between 18–40 m (60–130 ft), the canopy has more life than any other layer. The leaves and branches of the canopy are so dense that they all seem to connect. There is plenty of habitat structure for arboreal animals, which spend their lives in the treetops. Everything an animal needs is provided by the canopy. There is ample food and shelter, so some creatures never venture to the ground. Bananas, avocados, and guavas are some of the fruit that abundantly grows under the canopy. Consumers such as bats, sloths, and monkeys become prey for larger consumers.

Thirty percent of all of the world's bird species, such as macaws, toucans, kingfishers, hummingbirds, and owls, live here. Geckos and snakes camouflage to match their surroundings and stay better hidden from predators or to sneak up on prey. Suction cups on the bottom of the feet of red-eyed tree frogs keep them vertical without falling, and chameleons, with the ability to move each eye separately, are always on guard.

arboreal (adj.)

Arboreal animals live mainly in trees.

Because there isn't much wind, trees rely on animals to carry seeds and pollen. The sword-billed hummingbird uses its long bill to reach deep into a flower. The bill gets coated with pollen, which gets carried to the next flower. Strong smells and dazzling colors of plants attract sloths, monkeys, and other animals. As seeded fruits are eaten, the seeds are dropped and dispersed in the animals' waste.

There is plenty of sunlight and water here, so many plants grow. A lot of plants have drip tips, which are long, narrow ends that let extra water drip off. This helps the plants stay dry.

Notable Creatures of the Canopy

Silver Vase Plant

This bromeliad is native to Brazil. As an epiphyte, it grows on other objects such as rocks, logs, or tree branches by attaching its roots. Since the roots are not in soil, it extracts nutrients from leaves or pieces of wood that have fallen and been caught in its leaves and petals. The upturned shape of the plant parts allows water to trickle into the plant's body. The pools of water that collect are also good locations for dart frogs to deposit their tadpoles. This plant has only shoots and new stems. After a shoot blooms, it dies.

Emerald Tree Boa

Mimicking the look of a large vine by taking on a vibrant green camouflage, the emerald tree boa remains undiscovered not only by its prey but also by its predators. This giant snake grows to an average of 1.8 m (6 ft). Parrots, monkeys, and other canopy-dwelling animals are prey to this sneaky creature. Since it does not produce venom, the boa traps its prey by coiling around them. Like monkeys, the emerald tree boa has a prehensile tail allowing it to maintain a firm grip around branches.

Margay

Easily confused for an ocelot, margays are much smaller and more dominant than other creatures of the canopy. To aid in balance, margays have long, heavy tails. They have soft, broad feet, agile toes, and ankles that allow them to turn their hind feet backward 180 degrees. Having such flexible feet enables them to hang from the branch of a tree with only one hindfoot. They can propel through trees quickly, and if they happen to slip, they can easily catch a branch with only one hindfoot and continue their climb.

The Emergent Layer

Towering above the canopy, you've entered the emergent layer, which reaches about 50–60 m (160–200 ft) in height. The sun beats down upon your neck, and your hat catches in the wind. The animals that live here are tough and resilient. Some of the same birds, bats, and insects you saw at the canopy will come to this layer only if they are skilled enough to live in this part of the rainforest. Here, the branches are more fragile, so it is the animals that are lightweight or are good flyers, gliders, or climbers that can dwell here. Their presence in this layer is essential as they help pollinate the flowers of the tallest trees. Some creatures found in the emergent layer include scarlet macaws, blue-and-yellow macaws, harpy eagles, howler monkeys, spider monkeys, iguanas, and gliding tree frogs. To eat, one must climb back down to the canopy for a meal and then return to the emergent layer.

The trunks of emergent trees are often bare from the base through the midsection; leaves and flowers only appear at the top. These trees are more rugged, having waxy leaves that tolerate the sun and wind and retain water. This is especially necessary during the dry season. Trees in this layer produce lightweight seeds, some suited with little wings for better seed dispersal so that the wind can carry them. Remember that agouti chomping away at the Brazil nut? That nut fell from the top of a Brazil nut tree that is found in this layer. Because of the distance of the fall, the nut can travel as fast as 80 kph (50 mph) before it hits the ground! Also, remember those strong and sturdy buttress roots? They hold up another giant tree found here called the kapok [KAY–pock] tree.

Creatures of the Emergent Layer

Blue Morpho Butterfly

With a striking bright blue hue on the top side of the wings and a camouflaged brown shade on the underside, these morpho butterflies both catch the attention of others and hide very well. Rippled through the blue side are microscopic scales that reflect light, giving the butterfly an iridescent glow. The blue on a male morpho butterfly is more vibrant, which attracts females. When the wings are closed, the brown underside uses mimicry as found in "eye spots" which scare off predators. As these creatures fly, the flicker between the dark and bright wing sides makes them look as if they're disappearing and then reappearing. Blue morphos are one of the largest butterflies in the world, with a wingspan of about 20 cm (8 in).

Spider Monkey

Swinging with ease through the top branches of the emergent layer, these monkeys truly look like spiders when hanging upside down. Their prehensile tail is essential, as it can grasp on to branches firmly. Spider monkeys freely hang while their arms and legs are foraging or playing with other members of their group. Baby spider monkeys start using their tails early as they wrap them around their mother's tail while riding on her back. Spider monkeys have long fingers and no thumbs. These large primates can weigh up to 10 kg (22 lbs).

Kapok Tree

Kapok trees tower over the sea of leaves and rise into the emergent layer. Kapok leaves are small and more narrow, which helps reduce the amount of water that transpires out of them. Their seeds are quite unusual. Within a pod-shaped fruit lays a web of white cotton-like fibers and black seeds. When the fruit opens to release seed, the wind catches the fiber, helping with seed distribution. The flowers of the kapok tree bloom at night and offer an unpleasant smell, actually attracting bats, which are the primary pollinators for this tree.

Birds of the Amazon

The Amazon is home to about 1,300 species of birds, and they come in an assortment of sizes, shapes, and colors. Echoes of melodious tunes, squawks, and whistles formed by birds ring throughout the forest. Beaks can be pointed, long, wide, hooked, or flat, each kind specialized for that bird's way of feeding. Hummingbirds feed on nectar, toucans saw open fruit, woodpeckers drill away at tree trunks to snag insect larvae, and eagles and hawks hunt for small mammals. Plumage, wings, and tailfeathers can be found in all colors of the rainbow. Each type of bird is unique in its design.

Amazon Kingfisher

These tiny yet stocky birds have beaks that seem a bit too big for their bodies, but these beaks come in handy for catching prey. Both the male and female work together to build their nest by digging out a tunnel with their feet to hide their eggs. These birds like to stay clean by diving into water and then perching to preen.

King Vulture

These carnivores feed on carrion (decaying animals). King vultures are truly treated as kings. They are granted first eating rights by smaller vultures with stronger beaks that prepare the carcass for eating. King vultures sport a distinguishing caruncle on their faces (the fleshy wattle), but its purpose is unknown.

Crimson Topaz

Residing high in the canopy are these beautiful hummingbirds. Their striking plumage has a variety of crimson, bronze, yellow, and green with a metallic sheen. Their beauty is used to attract females and assert dominance over territory. These dainty birds feed on nectar from flowers and also catch insects.

Hyacinth Macaw

Macaws are one of the most common birds in the Amazon, and these brilliant blue hyacinth macaws are the largest of them all. They can live for up to 50 years, which they spend with a lifetime mate. Groups of these birds can be found eating clay at banks along the Amazon basin called macaw clay licks.

Symbiotic Relationships

As you journey through the rainforest, you discover interesting interactions between different creatures. Many of these interactions we are witnessing are symbiotic relationships. A symbiotic relationship is formed between two animals when they benefit each other in some way. We've seen this with the agouti and the Brazil nut tree and with the leafcutter ant and the fungus that grows on the leaves.

A Frog and a Spider

Small dotted humming frogs and certain types of burrowing tarantulas have an unusual relationship. Typically, frogs make easy prey for these tarantulas, but not the dotted humming frog. Special hairs on the legs of the tarantula sense chemicals, and when those hairs graze against this frog, they tell the spider not to eat it. The frog then secures a safe place to dwell by sticking around in the spider's burrow, feasting on ants that would typically eat the spider's eggs and spiderlings.

Tree Roots and a Fungus

An important, yet easily unnoticed, player in the ecosystem is a special kind of fungus called mycorrhizae [mai-kuh-RAI-zee] that lives on the unique root systems of forest trees. Since most nutrients are found in the decaying leaf litter and not in the soil, the fungus helps the roots absorb those nutrients. In return, the fungus is given shelter and sugars that the plants produce.

Resources in the Amazon

The Amazon Rainforest serves a greater purpose than merely exquisite beauty. For people native to the Amazon, their families have depended on the food, materials, and medicines provided by the rainforest for many generations. Similar to the way many people travel on roads, the people of the Amazon travel on canoes through the rivers. They are in balance with nature, taking moderately to fill their needs and protecting and respecting the land. However, the Amazon doesn't provide only for those that live nearby; its providence reaches people worldwide.

Primarily, as the biggest area of dense vegetation, the Amazon absorbs much of the solar radiation that hits our earth and recycles vast quantities of carbon dioxide for oxygen.

Food, medicine, and wood from the rainforest are used by many throughout the world. Rubber, bamboo, peanuts, sugar, tapioca, mangoes, bananas, macadamia nuts, and coconut oil are on the list of the products people use from the Amazon. Twenty-five percent of western medicine comes from plants of the Amazon, and much of it is used to treat cancer. A compound found in rosy periwinkle, for example, is used to treat leukemia and Hodgkin's lymphoma. Malaria treatments come from a drug called quinine, which is derived from the Amazon's cinchona (suhn-CHOW-nuh) tree. Locally, the rivers supply fisheries and provide water for agriculture and hydroelectricity.

Rosy periwinkle

Conservation & Threats

When the bounty of the rainforest is not taken in moderation, the beautiful creatures—animals and plants— face many threats. A main concern for the Amazon is deforestation. Trees are cleared for logging, much of which is done illegally, or they are burned to clear the land for agriculture or ranching. As trees are destroyed, so are many habitats for the creatures that depend on the Amazon—the homes for the epiphytic plants, the nests for the harpy eagle, and chrysalides for the morpho butterfly, for example.

Once the land is cleared, it is very unlikely the rainforest will return. For example, removing trees also removes that special mycorrhizae fungus that allows nutrients to better attach to roots. Instead, the nutrients get washed away. In addition, when too many trees are removed at one time, the animals that were responsible for helping disperse seeds are unable to do so as their homes are destroyed. Rainforest destruction is alarming at this time due to its rapid rate. The lifecycles of these plants and animals depend on the rainforest being left as it is, and too much land is being cleared too fast.

Deforestation is not the only threat the Amazon faces. Illegal hunting and trafficking of animals, such as the wildcats and sloths, can drive these remarkable animals

Since 1970, deforestation has caused 700,000 sq km (270,300 sq mi) of the Amazon Rainforest to be lost.

What would happen if the entire rainforest was cleared?
More than 5,000,000 sq km (2,000,000 sq mi) of lush, green forest would become barren and suffer both dry spells and flooding. Carbon dioxide usually absorbed by the trees would enter our atmosphere. Millions of creatures would no longer have a home, including the people who depend on the Amazon.
With no plants rooting the soil in place, the soil would erode away. The world, once dependent on food and medicine from the Amazon, would no longer have those resources. Without the trees and the thick cloud layers, the water cycle would be disrupted and affect the world's climate.

to extinction. Illegal mining pollutes the water, on which 31 million people directly rely.

With this realization, many organizations and communities are working hard to educate people and employ sustainable practices so that the long-term impacts of deforestation are considered over the immediate benefits. Several South American countries that have borders within the Amazon are putting rainforest preserves into place. For example, Brazil has worked tirelessly to break these destructive patterns and has established the largest protected area. Many of these efforts to protect the land are being monitored by law enforcement and satellites, but it is still a difficult task as the Amazon is massive. With greater awareness and wisdom in how the Amazon is being used, there is hope for a future as bright as the beautiful rainforest life.

The Amazon: Featured Creatures Fun Facts

Emperor Tamarin

Habitat: Canopy
Ecological Role: Omnivore; help with seed dispersal as they eat fruit; prey for snakes, pumas, and other predators
Diet: Fruit, nectar, flowers, insects, gums (resin/sap)
Lifespan: 10–20 years
Threat: Not endangered but habitat is facing deforestation
Fun Fact: Males and some females have dichromatic vision (seeing the world in two colors) which helps them see through camouflage.

Capybara

Habitat: Forest floor and near water
Ecological Role: Herbivore; prey for jaguars, caimans, ocelots, harpy eagle, and anacondas
Diet: Vegetation
Lifespan: Up to 7 years
Threat: Not endangered but are hunted for their fur by some people
Fun Fact: Like all rodents, capybaras' teeth continually grow, so they must continue to gnaw on things to wear their teeth down.

Three-Toed Sloth

Habitat: Canopy
Ecological Role: Herbivore; prey for jaguars, snakes, and large birds
Diet: Vegetation
Lifespan: 25–30 years
Threat: Not endangered but habitat is facing deforestation
Fun Fact: Green algae grows on the fur of these sloths, which acts as a camouflage.

The Amazon: Featured Creatures Fun Facts

Howler Monkey

Habitat: Canopy
Ecological Role: Omnivore; prey for pumas, jaguars, harpy eagles, and snakes
Diet: Fruit, flowers, leaves, and nuts, occasionally bird eggs
Lifespan: 15–20 years
Threat: Not endangered but habitat is facing deforestation
Fun Fact: Their loud howl sounds more like a bark and is used to communicate territory; a cup-shaped hyoid bone in their throat amplifies the sound.

Pink River Dolphin

Habitat: Rivers and lakes
Ecological Role: Piscivore (carnivore that consumes primarily fish); prey for caimans, large snakes, and jaguars
Diet: Fish, crustaceans, and turtles
Lifespan: 10–30 years in captivity, unknown in the wild
Threat: Not endangered but river pollution and human activity associated with fishing are threats
Fun Fact: They are born gray and slowly become pink as they age; they "blush" more pink when excited.

Giant River Otter

Habitat: Rivers and streams
Ecological Role: Carnivore; top aquatic predator and helps control prey population; prey for jaguars, pumas, caimans, and anacondas
Diet: Fish, crabs, turtles, small caimans, and snakes
Lifespan: Up to 8 years
Threat: Endangered; habitat loss, river pollution, hunting, and fishing are threats
Fun Fact: They are sociable and live in groups with up to 20 other otters.